Text and illustration copyright © 2022 by KiwiCo, Inc.

All rights reserved. No part of this book may be reproduced in any form without written permission from the publisher.

ISBN: 978-1-956599-04-6

Library of Congress Control Number: 2021953222

Manufactured in China.

10 9 8 7 6 5 4 3 2 1

KiwiCo, Inc.
140 East Dana Street
Mountain View, CA 94041

KiwiCo.com

LOOK and WIGGLE

written by
Jen Malone

KiwiCo
Press

WIGGLE, WIGGLE.

Brooks had his first loose tooth! It was so satisfying to watch it jiggle.

But what if it hurt when it came out?

Or worse, what if he accidentally swallowed it?

His sister Rosie popped into the bathroom.
"Hey Brooks, Mom's waiting. We gotta go!"

She reached around him and drew a smiley face in the steam.

"I can't," Brooks said. "I have to keep an eye on my tooth!"

Rosie was already a total pro about losing teeth. "Don't worry—there will be mirrors everywhere we go. Promise!"

Mirrors everywhere? Brooks doubted that! But he closed his mouth and followed her out the door.

TRY THIS!

Steam is caused when warm air meets cold mirror. In science, this is known as *condensation*. In art, it's the perfect canvas for disappearing creations. Breathe on your mirror and draw whatever you like!

On the way to the vet with Mom and Brewer, they passed the science museum.

"See? Rosie said. "What did I tell you?" She pointed to tall mirrors sticking out on either side of a school bus.

"Those help me see behind me while I look at the road in front," the driver told them.

"Maybe Mom doesn't really have eyes in the back of her head," said Brooks. "Maybe she just has a mirror!"

TRY THIS!

Can you see the back of your head while facing forward? Hint: You'll need two mirrors!

A scientist from the museum saw them looking at the mirrors. "Did you know we use mirrors to see the entire universe?"

Brooks and Rosie gasped.

"It's true. The new Webb Telescope has an enormous mirror that can see billions of light-years away."

"What's a light-year?" asked Brooks.

"It's a distance of millions of miles," said the scientist.

FUN FACTS!

The mirror on the Webb Telescope is too big to fit on any space shuttle, so engineers had to make one that could fold up for transport into space, then flatten once there. It's six times bigger than the last telescope we used to peek into the deep unknown.

Did you know sunlight takes eight minutes to travel from the sun to Earth? Any time you see sunshine, you're actually seeing light from eight minutes ago, before it traveled across the universe to your eyes.

Finally, they arrived at the vet's office. Brewer got up on the table, and the vet pulled something from her pocket. At first, Brooks thought it was a lollipop, but it was—

"Another mirror!" Brooks exclaimed. Rosie grinned.

"This little mirror lets me peek into Brewer's ears," the vet said. "Your dentist uses one just like it to see the back of your mouth."

"I have a loose tooth." Brooks showed her. "But it's in the front, so you don't need a mirror."

"Sure don't," she said. "Hey, can you keep a secret?"

Brooks and Rosie nodded.

"Know what I especially love using this for?"

TRY THIS!

Calling all secret agents! Your mirror lets you peer around corners while you stay hidden.

As they walked Brewer home, Brooks wiggled his tooth and daydreamed about mirrors bigger than space shuttles.

"*Oof!*"

He crashed into their mailperson, Pat.

"Whoa there!" Pat said.

Brooks peered up at him, but all he could see in the lenses of Pat's sunglasses was . . .

. . . TWO OF HIMSELF!

"Your eyes are mirrors!" Brooks exclaimed.

Pat chuckled. "My eyes water without protection. These sunglasses help."

"How?" Rosie asked.

"The mirrored lenses in my glasses bounce all that sunlight shining in my face away from my peepers."

"And I can watch two of my loose teeth while we talk," Brooks said.

TRY THIS!

Clone yourself! Standing at a big mirror, angle your small mirror to try to create two (or more!) of you.

Brewer tugged at his leash.

Pat laughed. "Someone's ready to get home."

"But first, I have to drop these two off at class," Mom said.

"SCOUTS!"

Brooks cheered.

"ART!"

Rosie whooped.

They got to the community center early. While Brooks waited for his Scouts troop to start, he joined Rosie's art class.

"I've been waiting on a sunny day for today's fun theme: MIRRORS!" Mr Ming said.

Brooks elbowed Rosie. "Can you believe it?"

"Let's head outside!" Mr. Ming led the way.

"First," said Mr. Ming, "do not look directly into the sun or at the sun's reflection. That can hurt your eyes."

Mr. Ming placed his paper on the ground in a sunny spot, then put his mirror on top. "Now move the mirror around until you see . . ."

"A rainbow!" Rosie gasped.
"There's a rainbow on my paper."

TRY THIS!

Psst! You can also catch the rainbow. Find a sunny spot, and stand your mirror up straight on a piece of white paper. Now move your mirror slowly until you find a rainbow.

It was time for Brooks' to get to his scouts meeting. He reached out to return his mirror, and his fingers slipped!

"Yikes!" Mr. Ming cried, grabbing it before it could fall. "If you break a mirror, it's seven years of bad luck."

OH NO! thought Brooks. He couldn't risk any bad luck with his tooth so loose.

Mr. Ming winked. "Good thing it's only a superstition."

More Mirror Beliefs

Romans were the first to manufacture mirrors, and they believed their gods saw souls through them. Breaking one was disrespectful to the gods, who would punish the breaker with bad luck.

The Chinese practice of feng shui says that your body releases negative energy through sleep. To avoid having those bad vibes reflected onto you, avoid placing a mirror facing your bed.

In the Jewish faith, mirrors are covered after the death of a love one to keep the person's soul from becoming trapped.

At Scouts, Brooks joined his friends by Mr. Jefferson.

"Today we're learning survival skills for our camping trip next month," Mr. Jefferson said. "We probably won't need them, but good scouts are always prepared!"

He held up a small mirror. "I borrowed this from Mr. Ming. He uses it to find rainbows, but I can use it to find other people. I can signal a helicopter in the sky or a rescuer as many as seven miles away. Check out how."

1

Facing the sun, place the mirror next to your eye.

2

Stretch out your other arm in front of you and find the reflection of the mirror on your hand.

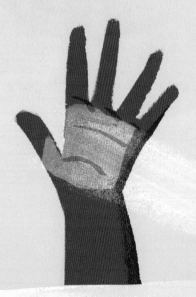

3

Spread open two fingers to create a V and guide the reflected light between them. Keep a small portion of the reflection visible on your fingers so you know it's there.

When they got home, it was time for dinner. Brooks and Rosie made a list of all the mirrors they'd encountered that day.

"Can you think of any more?" asked Mom.

Brooks had lots of ideas!

Brooks examined his loose tooth in the back of his spoon (mirrors really were everywhere!).

He checked between every mouthful, just to make sure he hadn't swallowed it.

He took a big bite of carrots, and then . . .

"MY TWOOFTH!"

The Tooth Fairy was coming tonight!

And Brooks was going to catch the Tooth Fairy. He positioned all the mirrors he could find around his room so he could see every corner from his cozy spot in bed.

"Just try to get by me, Tooth Fairy!"

Brooks carefully tucked his tooth under his pillow and snuggled under his covers. He was ready.

The morning sun hit the mirror over Brooks' dresser and aimed a beam right into his face.

Brooks woke up. He hadn't caught the Tooth Fairy!

In the spot where he'd left his tooth was a coin almost as reflective as a mirror. Next to the coin, he found a note.

Dear Brooks,
I love to play tricks, too.
What does a woolly, prehistoric creature use to chew its food?
Love,
The Tooth Fairy

Brooks flipped the paper over.
There was no answer, only . . .

Can you solve the tooth fairy's riddle for Brooks?
(Hint: Use your mirror!)